PRACTICAL ESSENTIAL BOOK FOR HVAC

Ultimate guide for installing A/C, heater and refrigerator maintenance

Kelvin C. Charles

Table of Contents

CHAPTER ONE

INTRODUCTION FOR HVACS

A heater is a unit that courses hot air all through a space to keep it warm. Heaters contrast from boilers along these lines, utilizing heat flow rather than water. A heater utilizes two kinds of starters to create heat.

OPPOSITION AND CONTRAST

Ignition for gas heaters and electric opposition for electric heaters; after the heater produces heat, the office's central air framework scatters it through different vents and channels and warms the structure.

DIFFERENT SORTS OF HEATERS

THE PROPANE HEATER

This sort of heater involves propane as a fuel. Once lighted, heat trade happens and hot air scatters into the vents of the house.

THE GAS HEATER

A gas heater is by a wide margin the most well known kind of heater. Gas heaters utilize flammable gas for burning to produce heat. Flammable gas travels through to the unit through an outside gas line. The far reaching utilization of gas heaters is because of the great measure of intensity that they put out. Homes situated in chilly districts by and large use gas heaters solely along these lines. Gas heaters are additionally incredibly practical.

THE ELECTRIC HEATER

An electric heater is adroitly not the same as different kinds of heaters. They utilize electric curls, called components, to make heat. A resistor interfaces power from the home or office into this heater and warms the approaching air.

Electric heaters furnish clients with elevated degrees of adaptability in warming.

PRACTICAL STEPS IN INSTALLING COOLING FRAMEWORK (A/C)

Off close

Begin setting down floor security and getting devices and hardware set up, existing climate control system contains a refrigerant that should be taken out before another one can be introduced. When the refrigerant is recuperated appropriately, electrical wiring is detached from the current forced air system.

Create convey

Close to the climate control system is a distinction box, which is a wellbeing gadget in the event of crises. It can close down your A/C unit. From the distinction box, a whip adaptable electric course conveys the capacity to the climate control system.

Set-up and lift

The current area of the forced air system will require some prep work before the new climate control system can be set up. This prep work incorporates supplanting the cushion that the new forced air system will set on.

When the new climate control system is introduced, it can't be moved or lifted securely without harm much of the time.

Loop surface

Moving a large number of its associated could for all time harm the framework. A composite cushion can stay away from harm to the climate control system on the off chance that it should be evened out from now on.

Current align

The indoor evaporator loop is the other portion of your cooling framework that sits over the heater, however now and again it will sit beneath the heater. This is the thing the opposite finish of the refrigerant lines is associated with.

Allow variety access

The indoor evaporator loop has two varieties. One is cased and the other is uncased. In one or the other application, the sheet metal plenum should be separated before the new loop can be introduced. This interaction can be convoluted when admittance to the loop is confined.

Place Refrigerant Lines

Then, the current refrigerant line set can be taken out. This is a bunch of two copper lines through which the refrigerant goes from the outside condenser, then back into the indoor evaporator loop. The refrigerant line set is normally tied to the lower part of the floor joists.

Curl off and board

These lashes are eliminated, and the line set is brought down. It's suggested that the line set be supplanted for each situation that it is uncovered and available. There are two choices while introducing a new evaporator curl for any new climate control system framework; cased and uncased. The favored choice is a cased curl, since it arrives in a protected bureau with removable boards on the front that permits admittance to the loop inside.

Make adjustment

The cased curl is additionally intended to sit on top of the heater with no alterations, which makes fixes and adjustments simple. Generally speaking, matching the shade of the furnace will be painted. While utilizing a cased curl, it is now introduced by the maker and just should be put on the highest point of the heater. The last step by then is to associate and seal it to the current sheet metal plenum.

Loop case and smooth

An uncased curl is the loop without a protected bureau. They have more establishment prerequisites and longer planning time. They are accessible in similar varieties as the cased curl. Uncased curls are normally harder to introduce. Notwithstanding, an uncased loop will give a similar solace and be similarly pretty much as effective as a cased curl when introduced appropriately.

CHAPTER TWO

AC AND MOUNTING PATTERN

HOW TO MOUNT AND REPLACE COPPER REFRIGERANT LINES

Differential

The new refrigerant line set is a mix of two individual copper lines. One line is generally greater than the other and is protected. The greater line is known as the attractions line, and the little one is known as the fluid line. They come bundled together, moved in a curl. It is accessible in numerous determinations of lengths, from 16-50 feet. You will seldom require in excess of 50 feet in a private home.

Twist and turn

The key is introducing the line with as couple of twists and turns as could really be expected. At the point when these twists are essential, the delicate copper can be bowed the hard way or a tubing drinking spree. In the event that you don't as expected safeguard the line, it can erode. Galvanic erosion additionally called bimetallic consumption is the breakdown of metal when two disparate metals come into contact. This requires some investment to happen.

Fluid line and pull

By the by, whenever permitted to happen, it could ultimately decrease the uprightness of the copper refrigerant line over the long run until it causes a hole.

The line set will run from the forced air system outside to the new indoor evaporator curl on top of the heater. After the attractions line has been introduced and gotten, the more modest, fluid line is run along the pull line, got in similar holders.

Crimps and control

A line set ought to be new, introduced as one piece, liberated from any sharp twists or crimps, and got with a fitting hanging framework. There will likewise be a little, low-voltage control wire that runs from the heater along the line set to the forced air system outside.

On and off

The little existing wire coming from the indoor regulator to the heater conveys a low voltage message to the climate control system to turn on and off when a call for cooling is required or has been fulfilled. Evened out it, and our refrigeration lines and electrical disengage are fit to be wired into the new forced air system condenser. Eliminating the new forced air system and setting it on the cushion.

IMPORTANCE OF CHANNEL DRYER INSTALLING

Channel driers are a vital part in the refrigerant framework. They serve two primary capabilities. The main capability of a channel drier is to ingest dampness. The second is to give actual filtration. While introducing another forced air system condenser, you should continuously introduce a fluid line channel drier into the refrigeration framework. Most new climate control system condensers accompany a fluid line channel drier for you to introduce.

EASY ESTABLISHMENT GUIDE ON HVACS

Set conduit

Ventilation work can likewise be produced using a wide range of materials, like aluminum, excited steel, tempered steel and even copper. In any case, the most widely recognized materials utilized in homes are aluminum and stirred steel. As well as being protected, ventilation work will likewise should be fixed off with pipe tape any place two bits of channel meet.

Cut opening

This will make an impermeable fit, which will work on the general proficiency of the framework's wind stream. You should associate your return line to the radiator by cutting an opening in the side of the warmer. Warmers are not made with precut openings since return conduits shift in size starting with one house then onto the next.

Power on

Make certain to actually look at your radiator's establishment manual to guarantee that you are interfacing the return conduit to the right side prior to cutting into the sheet metal outwardly of your warmer. Inside the lower part of the warmer, you will find a blower engine that maneuvers air into the radiator and powers it up to the inventory conduit.

Confront and range

Warmers accompany both left and right-confronting blowers to match the ventilation work of the home where the radiator is being introduced. Another way you can guarantee that the blower is confronting the right side is by eliminating the intro page of the radiator. In the event that your blower isn't confronting a similar side as your return pipe, you should trade it out for one more warmer with the right direction.

Pipe accordingly

Whenever you have found the right area for your return pipe, cut an opening in the side of the radiator that matches the size and state of your bring channel back. Appending your return pipe is genuinely clear and normally doesn't need specific apparatuses.

Seal and return

Essentially butt the edges of your return channel up to the opening and seal the association with pipe tape. Likewise, your return channel ought to have an initial where you can introduce another air channel occasionally.

Sealant

Make certain to embed a channel into this space that is sufficiently enormous to cover the opening prior to turning on the warmer totally. Then, you should associate your gas lines to your radiator. Your warmer's gas line associations should be fixed utilizing a specific sealant since the strings are not water or air proof all of the time. Apply your sealant to the strings on your gas line, and then connect an adaptable gas hose from your gas line valve to your radiator's gas valve.

Wires link

After safely associating your gas line to your radiator's gas valve, you should interface the electrical wires from your warmer to your switch. Your new warmer will have two wires rushing to the space where your switch will be introduced. Match the impartial white wires, and interface them utilizing a wire nut or another type of electrical association. Additionally, ensure the ground wire green is associated with the ground screw inside the switch box.

Box weight

When you're unbiased and ground wires are associated, join the live wire dark to your switch by folding it over the screw on the switch, then; at that point, do likewise with the dark wire from the warmer. At the point when the switch is flipped, the circuit will be finished between the two dark wires, and your warmer will actually want to turn on.

Interface point

When your fundamental power supply is associated, you should interface the indoor regulator wires to your radiator's circuit board. After your electrical association is done, you should interface the ventilation work to the highest point of your radiator.

Upper and lower parts

Joining the ventilation work requires some customization, and you will need to gauge cautiously prior to slicing and putting the walls to your pipe association. There are a couple approaches to this; however the main piece of this step is guaranteeing an impenetrable fit. You will regularly interface the upper part with a piece of S channel, and the lower piece will be in a bad way straightforwardly into the highest point of the warmer or evaporator curl box.

Check, framework

Once your ventilation work is associated, close everything with conduit tape. After your radiator establishment is finished, you should check the gas and pneumatic stress for your framework. Prior to introducing your cooling framework, you should eliminate the board over the gas lines on the open air unit. The open air unit comprises of two essential parts.

Eliminate, packed

To start with, there is a condenser unit that packs and gathers the refrigerant. Second, there is a fan that helps eliminate the overabundance heat that is made when the refrigerant is packed. Inside the board over the gas valves, you will track down the electrical associations with power the unit. When your wiring is all set up, you should weld your gas lines. There will be two gas lines - one huge and one little. After you patch these to the relating gas valves on the outside unit, your open air establishment will be finished.

Pile and top

To introduce your evaporator curl, you should put it over the warming unit. On account of new development, your stockpile pipes ought to have proactively considered this, leaving you sufficient space to introduce your evaporator curl over your radiator.

Your evaporator will accompany a metal box that sits on top of your radiator, and this should be gotten on top of the warmer utilizing self-tapping screws.

Blend and tap

After it is in a bad way down, you should associate the highest point of the crate utilizing a comparable strategy to the one we depicted while looking at associating your radiator to the stockpile channel. You will need to close within joints utilizing a blend of conduit tape and channel sealant. In the wake of taping off the joints with channel tape, apply conduit sealant with a paintbrush and permit it to dry. Once your evaporator box is associated with the conduit, you can embed the evaporator loop and supplant the front board.

Weld and align

After your front board is supplanted, you can weld the gas lines that run from your open air unit to the evaporator loop. Now that everything is introduced, you can get back to the outside unit to open the gas valves, permitting the refrigerant to move all through the cooling framework. The framework can then be turned on, yet you might have to play out an extra tweaking to ensure the framework is running accurately.

HOW TO ENSURE SMOOTH USAGE OF HEATER SUPPORT

Power

Ensure the unit is getting power. Search for blown melds or stumbled circuit breakers at the principal entrance board. A few heaters have a different power entrance, normally situated at an alternate board close to the principal entrance board. A few heaters have wires mounted in or on the unit. On the off chance that the unit has a reset button, stamped reset and close to the engine lodging, stand by 30 minutes to let the engine cool, then press the button.

Minutes

On the off chance that the unit actually doesn't begin, stand by 35 minutes and press the reset button once more. Rehash something like again. On the off chance that the unit has a different power switch, do sure the switch is turned on. Check to ensure the indoor regulator is appropriately set. If vital, raise or, for a climate control system, lower the setting 5°.

Oil

In the event that the unit utilizes gas, check to ensure the gas supply is turned on and the pilot light is lit. Assuming it utilizes oil, check to ensure there is a satisfactory stockpile of oil.

SIGNIFICANT AND NEED OF MAINTENANCE OF A/C

Prior to accomplishing any work on a warming or cooling framework, ensure all capacity to the framework is switched off. At the really electrical entry board, trip the electrical switch or eliminate the circuit that controls the capacity to the unit. In the event that you don't know which circuit the framework is on, eliminate the principal wire or excursion the primary electrical switch to remove all capacity to the house.

Circuits

A few heaters have a different power entrance, for the most part at an alternate board close to the fundamental entry board. On the off chance that a different board is available, eliminate the wire or outing the breaker there. On the off chance that the wire blows or the circuit trips over and over when the heater or climate control system turns on, there is an issue in the electrical framework. For this situation, don't attempt to fix the heater. In the event that the unit utilizes gas and there is a smell of gas in your home, don't attempt to stop the gas or turn any lights on or off.